T5-BZD-464

READING POWER

Man-Made Disasters

Sludge and Slime

Oil Spills in Our World

August Greeley

The Rosen Publishing Group's
PowerKids Press™
New York

Published in 2003 by The Rosen Publishing Group, Inc.
29 East 21st Street, New York, NY 10010

First Edition

Book Design: Christopher Logan

Photo Credits: Cover, pp. 9, 16 (bottom), 18–19 © AP/Wide World Photos; pp. 4–5 © Bettmann/Corbis; pp. 4, 9, 12, 17 (globe) © PhotoDisc; p. 5 (inset) © Todd Gipstein/Corbis; pp. 6, 10 (maps and illustrations) Christopher Logan; pp. 7, 21 (inset) © Joel W. Rogers/Corbis; p. 7 (inset) © Larry Lee Photography/Corbis; p. 8 © Craig Aurness/Corbis; pp. 11, 12, 13 (bottom), 14, 15 © Natalie Fobes/Corbis; p. 13 (top) © Alan Levenson/TimePix; pp. 16 (top), 18 (inset) © Gary Braasch/Corbis; p. 17 © AFP/Corbis; pp. 20–21 © AP/Wide World Photos, courtesy Phillips Alaska Inc.

Library of Congress Cataloging-in-Publication Data

Greeley, August.
Sludge and slime : oil spills in our world / August Greeley.
p. cm. — (Man-made disasters)
Summary: Explores what oil is and how it is used, the harm that is done when it is spilled into the ocean, and what can be done in the future to prevent such disasters as the 1989 wreck of the Exxon Valdez.
Includes bibliographical references and index.
ISBN 0-8239-6485-X (library binding)
1. Oil spills—Environmental aspects—Juvenile literature. [1. Oil spills—Environmental aspects.] I. Title. II. Series.
TD196.P4 G74 2003
363.738'2—dc21

2002000511

Contents

Oil

Oil is used for many things. It is used to heat homes and to make gasoline for cars.

Check It Out

Each day, the United States uses over 700 million gallons of oil. The whole world uses almost 3 billion gallons of oil every day.

Oil burns very easily.

Oil is even used to make crayons, candles, and plastic. However, oil can be very harmful if it spills.

Many of the things we use every day are made from oil.

Oil is a greasy liquid that will not mix with water. It is found deep beneath the surface of the earth. Millions of years ago, layers of mud and sand covered dead plants and animals. As the years passed, pressure and heat from the mud and sand turned the dead plants and animals into oil. People dig deep in the earth to get oil.

From Plants and Animals to Oil

Oil can be made on land or in the ocean. In the ocean, plants and animals die and fall to the bottom.

Layers of mud and sand build up on top of the dead plants and animals.

Over time, pressure and heat from the mud and sand turn the dead plants and animals into oil.

Heavy machines, called rigs, are used for drilling into the earth to get oil.

Oil is shipped to places all over the world. Big ships, called tankers, carry the oil. Sometimes, accidents happen and the tankers spill oil.

Oil tankers are very large. Some tankers are over 1,500 feet long.

Check It Out

In the United States, about 14,000 oil spills happen each year. These spills happen on land and water and are both big and small.

Spills can be caused in many different ways. This tanker ran into another ship, causing more than 1,000 gallons of oil to spill.

The *Exxon Valdez*

On the night of March 24, 1989, the tanker *Exxon Valdez* left Alaska with its tanks full of oil.

As it neared ocean waters, the *Exxon Valdez* crashed into a reef. A hole was made in the ship. Oil started spilling out of the tanker.

The Exxon Valdez *oil spill was the largest spill ever in the United States.*

Nearly 11 million gallons of oil spilled from the tanker. The oil began to spread on the water. Thousands of people had to work fast to clean up the oil spill.

Check It Out

Every 22 minutes, the United States uses the same amount of oil that spilled from the *Exxon Valdez*.

The waves from the water washed the oil onto the shore.

Even so, the oil continued to spread for about two months, covering 460 miles of water. The oil on the water also washed up onto about 1,300 miles of shoreline.

Many people worked to clean up the oil.

The Cleanup

Fire hoses were used to wash the oil off of the beaches. Then, the oil was sucked up with special vacuums or wiped up with special sponges.

The oil cleanup took more than four summers. In the winter, when it was too cold for people to work on the cleanup, waves from winter storms helped wash oil off the beaches.

Bacteria that eats oil was also put on the beaches to help clean them up.

People cleaned the oil off of rocks by hand.

Many different kinds of plants and animals lived where the oil was spilled. Oil covered the feathers of birds and the fur coats of sea otters.

People worked to save as many animals as they could.

Oil stops an animal's fur or feathers from keeping the animal warm. Many people helped to clean oil off of animals so that the animals would not freeze to death.

Check It Out

Dishwashing soap worked the best to clean oil off of animals.

The Effects of Toxic Oil

Oil is toxic. Many animals died by swallowing oil or breathing in oil fumes. Food and water supplies and vacation areas were polluted and could not be used.

Scientists believe hundreds of thousands of animals died from the spill.

Many fishermen and other people lost their jobs.

Numbers of Animals Killed by the *Exxon Valdez* Oil Spill

Type of Animal	Number Killed
Seabirds	250,000
Sea Otters	2,800
Harbor Seals	300
Bald Eagles	250
Killer Whales	22
Salmon and Herring Eggs	Billions

After the *Exxon Valdez*

After the *Exxon Valdez* oil spill, the U.S. government worked to make sure that an oil spill like it would not happen again. The Oil Pollution Act of 1990 was passed. This law says that all tankers in the United States must be built stronger by the year 2015.

This is the first boat built stronger since the Oil Pollution Act of 1990. It has two layers of metal on its body instead of one. The extra layer makes it harder for oil to spill out of the ship.

The law also says that the owners of the tankers must have a plan in case of a spill. Hopefully, there will never be another spill as harmful as the *Exxon Valdez* disaster.

There is still oil on some beaches in Alaska.

Glossary

accidents (**ak**-suh-duhnts) unplanned events or happenings

bacteria (bak-**tihr**-ee-uh) very tiny living things; some can be useful, while others cause sickness

disaster (duh-**zas**-tuhr) a sudden event that causes great loss or harm

layer (**lay**-uhr) one thickness or level of something that is on top of another

pollution (puh-**loo**-shuhn) anything that dirties or poisons an environment

pressure (**prehsh**-uhr) the continued action of a weight or force

reef (**reef**) a narrow chain of rocks near the surface of the water

shoreline (**shor**-lyn) the place where a body of water and the shore meet

sponges (**spuhn**-juhz) special diaper-like materials made to suck up oil

tanker (**tang**-kuhr) a ship with tanks for carrying oil or other liquids

toxic (**tahk**-sihk) harmful or poisonous

vacuums (**vak**-yoomz) machines that clean by sucking up things

Resources

Books

After the Spill: The Exxon Valdez Disaster, Then and Now
by Sandra Markle
Walker & Company (1999)

Sea Otter Rescue: The Aftermath of an Oil Spill
by Roland Smith
Penguin Putnam Books for Young Readers (1999)

Web Sites

Due to the changing nature of Internet links, PowerKids Press has developed an online list of Web sites related to the subjects of this book. This site is updated regularly. Please use this link to access the list:

http://www.powerkidslinks.com/mmd/slsl/

Index

Word Count: 524

Note to Librarians, Teachers, and Parents

If reading is a challenge, Reading Power is a solution! Reading Power is perfect for readers who want high-interest subject matter at an accessible reading level. These fact-filled, photo-illustrated books are designed for readers who want straightforward vocabulary, engaging topics, and a manageable reading experience. With clear picture/text correspondence, leveled Reading Power books put the reader in charge. Now readers have the power to get the information they want and the skills they need in a user-friendly format.